# Cast in Stone
## The Molly Brown House Revealed

BY ELIZABETH OWEN WALKER

PHOTOGRAPHY BY JEFF PADRICK

## ACKNOWLEDGMENTS

To all the "K's" in my life: I would like to thank Kathleen Brooker for her support, interest, and critical eye throughout the process. Kris Christensen, thank you for shepherding us along, and for your attention to detail. To Kristine Hoehn, AIA, for her unfailing devotion to this property and for the technical expertise that helped develop the Preface. I would also like to thank the staff at the State Historical Fund for their reviews and assistance.

This project was collaborative from the beginning, but a few key players made it a reality. Especially Violet Carlon, whose spectacular and inspired design has left me breathless. Jeff Padrick for his incredible eye—Jeff, you made the house appear even more beautiful than I thought possible. Thank you Elizabeth Walker for diving into the research, writing such graceful prose, and accepting my sometimes not-too-subtle direction. I would also like to thank the Molly Brown House Museum staff, who supported Elizabeth and me while we were writing, meeting, and plotting. Finally, to Steve Grinstead, thanks for your gentle project direction, skill in editing, and your infinite patience with all of us!

Leigh A. Grinstead

I would like to thank the staff and volunteers at the Molly Brown House Museum, who provided me with the support and time to research and write this book. I also want to thank my husband, Frank, who encouraged me to write about what I love.

Elizabeth Owen Walker

This project was partially funded by a State Historical Fund grant award from the Colorado Historical Society.

*Facing page:* Photo courtesy Denver Public Library, Western History Collection.

Edited by Steve Grinstead
Designed by Violet Carlon
Printed in Denver by A. B. Hirschfeld Press

Text © 2001 by Historic Denver, Inc. All rights reserved.
Photographs © 2001 by Jeff Padrick.
All historical images are from the collections of Historic Denver, Inc., unless otherwise indicated.

International Standard Book Number: 0-914248-40-5

Historic Denver, Inc.
1536 Wynkoop Street, Suite 400A
Denver, CO 80202
(303) 534-5288
www.historicdenver.org

# MOLLY BROWN
## COLORADO'S UNSINKABLE HEROINE

She's the subject of books, movies, and a stage production. We know her as the "Unsinkable Molly Brown," because she survived the sinking of the great ocean liner RMS *Titanic*. What we think we know about her is largely unflattering: She was illiterate and unrefined; her husband, tired of her boorish ways, abandoned her; Denver's high society shunned her; and she shamelessly promoted herself, even taking advantage of the *Titanic* disaster to gain acceptance among the world's rich and famous.

Some of this is exaggerated. Most of it isn't even true.

*In the photo shown here, Maggie strikes a conversational pose. Wearing a Battenburg lace gown, she fairly exudes wealth and power, but perhaps most, intelligence. Life fascinates her; and surely, she will attract attention.*

Margaret "Maggie" Tobin was born July 18, 1867, in Hannibal, Missouri. Her parents were immigrant Irish, poor but by no means destitute. Margaret, like her brothers and sisters, attended grammar school, but the family's fragile finances dictated that the Tobin youngsters find work at an early age. Maggie, however, completed an eighth-grade education—no mean accomplishment in those difficult post–Civil War years.

At age thirteen she went to work, first in a tobacco factory, then as a waitress in a Hannibal hotel, where, folklore has it, she met the town's most famous son, Mark Twain. But Maggie would not stay in Hannibal forever. Her half-sister Mary Ann Landrigan had moved to Leadville, Colorado, a storied boomtown where silver kings were said to tumble over one another looking for ways to spend their newfound wealth. In 1886 Maggie joined the Landrigans, landing a job with the dry goods firm of Daniels, Fisher & Smith, where she sewed carpets and draperies. A fellow worker remembered Maggie as "exceptionally bright, a most interesting conversationalist, had a charming personality and this coupled with her beauty made her a very attractive woman."

Maggie had hardly unpacked before she met an upward-bound young miner named James J. Brown. Like Maggie, J. J. was Irish Catholic, and he was an experienced hand on the Colorado mining frontier. By 1886 he had risen from a common miner to shift boss, timberman, foreman, and mine manager. He was tall and handsome, and although not a silver king, certainly a man of some substance.

J. J. and Maggie were married on September 1, 1886, and by 1889 they had two children, Lawrence Palmer and Catherine Ellen, known as Helen. The newlyweds had settled in a comfortable, well-furnished home at 300 East Seventh Street. Maggie would remember these years as the happiest of her life. Her husband continued to prosper, she enjoyed the love of her children, and as a wonderful bonus, her parents and brothers and sisters now lived in Leadville.

These years were good to J. J., too. By 1892 he sat on the board of the Ibex Mining Company. The great Silver Crash of 1893 hardly affected him, for one of the Ibex properties, the Little Jonny silver mine, struck high-grade copper and gold. In the midst of financial ruin, J. J. and his fellow Ibex owners found themselves very, very rich.

Maggie and J. J. moved to 1340 Pennsylvania Avenue (later Pennsylvania Street) in Denver's plush Capitol Hill neighborhood, home to the Rocky Mountain West's greatest mining kings. Here, Maggie set about to conquer Denver's exclusive social elite. But she knew she had much to learn. "Perhaps no woman in society has ever spent more time or money becoming 'civilized' than has Mrs. Brown," opined *The Denver Post*. Maggie enrolled in New York City's Carnegie Institute, where she studied languages, literature, and drama.

A self-proclaimed "glutton for knowledge," she attended music lessons, schooled herself in great operas and symphonies, and took up culinary arts.

All this bore spectacular results. Maggie mastered French and German, as well as three or four other languages. The Browns gained entree to the city's social directory, and Maggie became a member of the Denver Women's Club and the Denver Woman's Press Club. She gave generously to Leadville's St. Vincent's Orphanage and to Denver's St. Joseph Hospital and the Cathedral of the Immaculate Conception. She also worked as a volunteer for Judge Ben Lindsey's juvenile court.

She received national attention after the world-shocking *Titanic* disaster. That night–April 15, 1912–Maggie found herself in a lifeboat. The one male occupant, the ship's quartermaster, panicked: "He was shivering like an aspen," Maggie remembered, "telling us our task in rowing away from the sinking ship was futile, as she was so large that in sinking she would take everything for miles around down with her suction."

Maggie took charge. Ignoring the sailor, she began rowing and instructed the others to do the same. She heard gunshots from the sinking liner, after which, wrote Maggie, there appeared "a rift in the water, the sea opened up and the surface foamed like giant arms spread around the ship, and the vessel disappeared from sight, and not a sound was heard." On board the rescue ship *Carpathia*, Maggie's linguistic skills enabled her to speak to the third-class immigrant passengers and make lists of the survivors so their families could be alerted. She also helped raise more than $10,000 for destitute victims of the tragedy.

Maggie now was a hero, the "Unsinkable Mrs. Brown." Denver proudly claimed her as a native daughter, and newspapers around the country clamored for her story.

But all was not well in the Brown household. Maggie and J. J. had separated in 1909. J. J. had entered an affair with a married woman, whose husband countered by suing him for alienation of affection. Maggie and J. J. went their own ways–Maggie to Europe, J. J. to mining properties in Arizona, Utah, and California.

Maggie turned to politics, fighting for maritime reform, and in 1914 running unsuccessfully for the U.S. Senate under the banner of the National Women's Party. During World War I, she entertained troops by acting in roles made famous by Sarah Bernhardt. For these activities, she was named to the French Legion of Honor.

J. J. died in 1922. Now, Maggie traveled in earnest, spending much of her time at European and East Coast resorts. A reporter wrote: "Heads turned at the sight of her, and as she passed by, she left in her wake an essence of violets, rose-water and mothballs. Everything about her was amazing and fascinating . . . . I sensed that she was a brave and lonely woman probably living in the heyday of her past."

Always there were things to do and causes to adopt: In her hometown of Hannibal, she dedicated a statue to Mark Twain; in Denver she purchased Eugene Field's home so that it could be preserved in Washington Park; in Paris, rumors circulated of her pending marriage to the Duke of Chartre; and during the tragic 1914 coal strikes in Ludlow, Colorado, she spoke out on behalf of the striking miners.

Toward the end, largely estranged from her children, she and her nurse lived at New York City's Barbizon Hotel. Here she died on October 26, 1932. Her death made headlines, but one Denver newspaper captured her life best: "She was a definite, fearless personality. She knew what she wanted and went after it, and seldom failed of her goal." *By David Fridtjof Halaas, Chief Historian, Colorado Historical Society*

*Adapted from* Colorado History NOW, *January 1998, published by the Colorado Historical Society. Maggie Brown's exceptional life is colorfully and accurately chronicled at the* **Molly Brown House Museum** *at 1340 Pennsylvania Street, Maggie's original Denver home. It is now owned and operated by Historic Denver, Inc., the city's only private preservation organization.* **For information, call the Molly Brown House Museum at 303-832-4092.**

# FOREWORD

A house reflects its owner's personality and soul. Nowhere is this truer than in the home of Margaret Tobin Brown, known to millions as "The Unsinkable Molly Brown." The Hollywood legend is not nearly as fascinating as Margaret's real life—a life and spirit that have proved inspirational for generations. Margaret was a "renaissance" woman, ahead of her time in many ways, and in these rooms you'll discover delightful evidence of her interest in travel, literature, and the arts. You'll also find hints of her philanthropy, her politics, and the deep passion she felt for humanitarian issues such as children's rights, miners' rights, and voting rights for women.

The child of Irish immigrant parents, Margaret grew up to become a woman who would entertain royalty, meet presidents, and even run for the Senate. Her home in Denver was the cornerstone of her business and social life—an elegant statement of a woman both practical and elegant, eclectic and refined. Margaret's friends included Denver society, European royalty, New York feminists, and heads of state. But she also remained close to the children of Denver's poorer families, survivors of the *Titanic*, and her old friends in Leadville and Hannibal. People from all walks of life came to enjoy the Browns' hospitality on Pennsylvania Street. Here, too, you'll discover loving reminders of the importance of Margaret's family. She raised five children in this home—a son and daughter and, when their mother died, her three young nieces.

Margaret lived through the ups and downs of Colorado's mining booms as well as the early years of the Depression. She enjoyed some of the best things this world has to offer and endured many tragedies as well. "Money can't make man or woman," she said. "It's not who you are, nor what you have, but what you are that counts." In these hallways, Margaret's indomitable spirit still lingers. I hope you carry a little bit of that spirit with you.

Kristen Iversen

Author, *Molly Brown: Unraveling the Myth*

Isaac and Mary Large commissioned William Lang *(facing page)* to design the house at 1340 Pennsylvania Street in 1886. Victims of the 1893 silver crash, the Larges needed to sell. J.J. and Margaret Brown, who lived in a modest home on York Street, worked out a simple arrangement with the Larges: They traded houses. Lang also designed the Everts house next door *(facing page, bottom)*. Architects often designed two, three, or even more adjacent homes for the same speculator. Photos courtesy Denver Public Library, Western History Collection.

The Browns and later owners replaced the front porch's original wood railings and balusters *(see above center)* with sandstone, extended the deck, and gave it a solid stone wall for support. The Browns converted two wooden back porches into an enclosed porch of brick and concrete. In their last major changes, they converted the drainage system and wood-shingle roof to red clay tile. Note the original wood porch at the rear of the home. Photo courtesy Denver Public Library, Western History Collection.

# PREFACE

𝒯ime and 𝒫lace  The Molly Brown House stands as an enduring symbol of its era and its city.

Away from the pollution, noise, gambling dens, and even seedier locales of downtown, Denver's Capitol Hill gave the privileged an added luxury: a grand view of the Rockies. In the 1880s the lucky few who made millions from the mountains, the railroads, or trade moved onto the Hill to display their newfound wealth. The result was a building and real estate boom in the neighborhood.

Between 1891 and 1894, mansions mirroring the exuberance of the Victorian era shot up throughout Capitol Hill. Well-known architects like William Lang, Frank Edbrooke, Fisher and Fisher, and Varian and Sterner designed homes both beautiful and elaborate, favoring the architectural styles of the day: the now-classic Queen Anne, the eclectic Richardsonian Romanesque, the refined Neoclassical.

But the boom couldn't last—1893 brought widespread economic panic. It was a nationwide depression, but Denver was hit especially hard. As one response to the panic, the federal government repealed the Sherman Silver Purchase Act. Under this act, the government had bought up silver to use as backing for U.S. currency. The act's repeal changed the money standard to gold—and in Colorado, silver mining had been king.

That same year, J. J. and Margaret Brown, then living at 1152 York Street, had the opportunity to buy the William Lang–designed house at 1340 Pennsylvania. J. J. had struck gold and copper as superintendent of the Little Jonny mine in Leadville. With gold's value soaring, the Browns found themselves exceedingly wealthy. They paid off their mortgage in cash the next year.

ℱorm and ℱashion  The Molly Brown House is a Queen Anne structure: Its roofline and asymmetrical facade typify that style, one of Denver's most popular of the day. Like its many projecting features (the house has another porch and two balconies), the house's ornamental wood panels, large decorative barge boards, and curved brackets are all classic Queen Anne elements. Capping the house, and pierced by five stone chimneys, is a tall roof with a lively arrangement of gables and hips.

But the house also has elements of Richardsonian Romanesque, a style characterized by rugged stone with intertwining decorative stonework. Durable Castle Rock rhyolite—the home's primary material—was known for its superiority. Easily cut to a consistent size, it breaks with a curved or scallop-shaped fracture, giving surfaces an unusual, rough appearance.

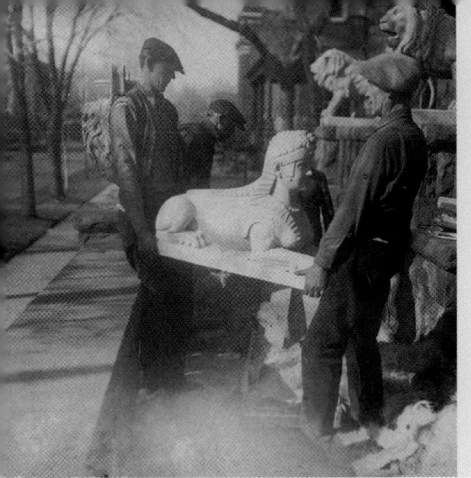

*Top left:* Workmen install the original sphinx statuary on the front porch, probably in the mid to late 1920s. *Left:* Looking south down Pennsylvania toward the Penn Garage; the garage still stands, now serving as loft and restaurant space. Courtesy Colorado Historical Society. *Right:* A view of Pennsylvania Street, showing the streetscape and housing demolished in the 1980s. *Facing page:* Lang's original floor plan for the Molly Brown House. Courtesy Denver Public Library, Western History Collection.

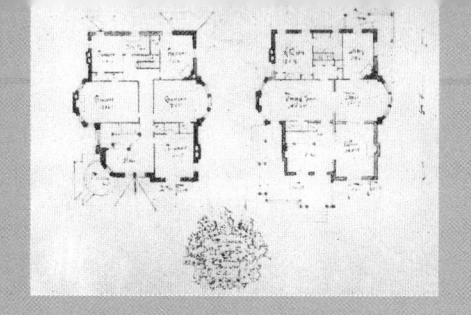

Lang complemented the rugged, dove-colored rhyolite with the smooth red sandstone that adorns the windows and front porch. The balusters, capstones, steps, and running trim are of sandstone as well. Like rhyolite, native Colorado sandstone was a favored building material. Probably quarried in Manitou Springs, most of the Molly Brown House's original sandstone was orange in hue, unlike the redder Lyons-area sandstone that replaced front porch elements in a 1980s restoration.

Built a bit later than the main house, the carriage house displays many of the same elements as the home. Lang directed that its main façade be constructed of Castle Rock rhyolite and sandstone trim to match the house. Its roof is clad with wood shingles and, like the house, it has terra-cotta elements at the ridge. A cupola caps the structure, and a door with flanking windows leads out to a small, decorative porch facing the house.

A stone retaining wall stretches along the sidewalk across the full width of the site. On it, a bas-relief plaque replicates part of a frieze located in Copenhagen and designed by the famous Danish sculptor, Theowaldsen. A reflection of Margaret's world travels, the plaque was set in the wall between 1904 and 1910.

## The Block

In 1893, four homes stood on the 1300 block of Pennsylvania. Lang had designed the house next door to the Browns' with a slightly smaller floor plan and of similar rough stone. The Browns' nearest neighbor was the Milheim house across the street. At the corner of Thirteenth was another Queen Anne. These were the block's sole residents until 1911, when St. Mary's Academy built its grand Neoclassical facility at the corner of Fourteenth.

With the silver crash, upper-middle-class houses went up alongside the more opulent mansions. Apartment buildings and modest homes housed servants and other working-class residents. In the 1920s, a wave of new apartment buildings flooded Capitol Hill. One such structure went up next door to Margaret Brown's home and with it the Penn Garage, a few doors south, in 1924.

When J. J. died in 1922, Margaret rented out rooms to supplement her income. With the onset of the Great Depression, only rarely was a house owned or occupied by a single family. In the years after Margaret's death in 1932, her home was no exception. Subsequent owners dramatically altered the house over the next thirty-eight years, creating more than twelve separate spaces for roomers.

## "Progress" and Preservation

In the 1950s and 1960s Denver underwent "urban renewal": Bulldozers demolished many of its finest buildings; apartment towers out of scale with the neighborhood went up; and houses came down for parking lots. Because the Brown house was still useful as a rooming house—and thanks to the sensitivity of its owner, Art Leisenring—it dodged demolition.

*Top left:* Mary Jane Johnson, a "roomer" who lived in the house. This January 1949 photo shows a stylish Mary Jane perched atop one of the lions. Visible in the background are the south side of the house and the porch roofline. The Molly Brown House was always a "rooming house," never a boarding house: There was no central kitchen, and meals were never served. Dorothy Weiss Wilson, a past resident, says she and her fellow lodgers snuck hot plates inside and stashed food on the window ledges. Courtesy Dorothy Weiss Wilson. *Above center:* In the nineteen-teens Margaret rented her entire house to the well-to-do Cosgriffs until their own mansion was built at 800 Grant. Railroad magnate David Moffat later bought that home. The impending demolition of the "Moffat Mansion" acted as an impetus for the founding of Historic Denver. Courtesy Denver Public Library, Western History Collection. *Top right:* An early Historic Denver volunteer—Arbor Day in the 1970s. *Bottom left:* Volunteers pose outside the Molly Brown carriage house amidst an early restoration. *Bottom right:* In the largest house move west of the Mississippi, the Milheim house is relocated to 1515 Race in 1989. Courtesy Ralph Heronema.

A growing nationwide movement led to the National Historic Preservation Act in 1966. As William J. Murtagh explains in *Keeping Time*, the act put specific measures in place for the evaluation of sites, buildings, structures, objects, and districts. Expanding the government's view of preservation, the act encompassed not just entities of national significance but those of state and local value as well—both historical and architectural. Locally, the city established the Denver Landmark Preservation Commission in 1967, allowing neighborhoods like Capitol Hill to designate historic districts in order to maintain the history of an area and ensure that development conformed to community standards.

In 1970, concerned citizens founded Historic Denver, Inc., to preserve the Molly Brown House. Restoration began the next year. Microscopic paint analysis, architectural research and survey work, studies into the historic and social context of Denver, and the removal of wallpaper layers to pinpoint period designs all led to a systematic interior restoration that has stood the test of time.

Preservation strides notwithstanding, the residential integrity of the 1300 block of Pennsylvania suffered. The destruction culminated with the removal of row houses and small-scale apartments to make way for a high-rise office tower across the street from the Molly Brown House. But with the creation of the Pennsylvania Street Historic District in 1996, the construction of high-rises and the demolition of homes are no longer permitted.

During its tenure as a rooming house, exterior changes accompanied the Molly Brown House's interior revamping. A fire escape, since removed, once served the current third-floor kitchen and a second-floor bedroom. Two windows were added on the north side: one in the kitchen and one on the second floor at a closet (once a bathroom). With the kitchen's restoration, the window was removed and the wall patched with matching rhyolite. The window in the closet is still visible from outside.

In 1998, a grant to Historic Denver, Inc., from the State Historical Fund set critical structural fixes in motion. Masons repaired, replaced, and stabilized brick and rhyolite in both the carriage house and main house. The Molly Brown House Museum has done everything from patching and sealing front porch tiles to making major drainage improvements to stabilize the entire structure. Further work has checked masonry deterioration on the front porch, brick deterioration in the basement, and damage to the carriage house's alley-front wall.

The grant has enabled the museum to interpret Margaret Brown's life while preserving her historically significant home through continued restoration. With staff offices moving out of the back parlor and back porch and joining the Museum Store inside the carriage house, the museum can now include the entire first and second floors of the main house in its tours.

These mammoth strides are protecting the Molly Brown House against the rigors of adaptive reuse in an ever-changing urban setting—casting in stone this indispensable structure's role to audiences worldwide.

Leigh A. Grinstead
Director, Molly Brown House Museum
Historic Denver, Inc.

*The Hatch Family,* 1870–71, Eastman Johnson (American). Oil on canvas. Courtesy The Metropolitan Museum of Art, Gift of Frederic H. Hatch, 1926 (26.97). Photograph © 1999 The Metropolitan Museum of Art.

VICTORIAN HOME LIFE

The Molly Brown House is emblematic of late nineteenth-century upper-middle-class homes, with such modern technologies as electricity, indoor plumbing, and central heat. Filled with the goods Margaret and J. J. Brown accumulated through travels and shopping, their home made a statement—to the outside world and to the Brown family itself.

The new middle class was modeling its houses on those of the wealthy. Architects designed homes in which public and private spaces were clearly defined. No longer multi-purpose, rooms served specific uses: The front parlor was for entertaining, the family parlor for every day. Usually, public spaces were near the front of the house, the utilitarian and private rooms at the rear. Public rooms conveyed status, while private rooms reflected the owners' more personal lives.

New technologies helped shape the rooms' spaces and functions. The advent—and, as time went on, the affordability—of central heating and electricity meant families no longer had to gather around a table under one lamp, or around a single stove. Separate heat sources allowed the dining room to become distinct from the kitchen and to develop its own uses and personality. Electric lights let family members enjoy various activities at the same time throughout the house.

Period literature described the home as a moral haven from the world's demands and a place to nurture children. Work, politics, and commerce were the world of men. The home was just the opposite: a miniature universe of culture and education for family and visitors. Women were charged with creating a home that communicated a family's status while providing that family with repose and moral uplift. The home's spaces reflected these beliefs. Now common in larger houses were "male" spaces like the Molly Brown House study and "female" spaces like the parlor.

Victorians' homes publicly displayed their owners' travels, status, and refinement—not to mention their cultural views, their education, and their wealth. Yet the home was a sanctuary from the outside world.

J.J., Margaret, Helen, and Larry Brown in Leadville, circa 1893. Courtesy the Colorado Historical Society.

*Left:* One of Margaret's blackamoor statues still greets visitors today. *Right:* The spindles of the staircase are machine carved—an example of the most up-to-date décor of the 1890s. *Bottom right:* The original tiles—a decorative touch in many Victorian homes—remain in the fireplace surround. These were made by the Trent Tile Company of Trenton, New Jersey.

### Restoration Note

The upper mantle of the fireplace was rescued from the William Lang–designed home next door before that house's destruction in the 1960s. Photographs show that it is very similar to this room's original mantle.

## THE ENTRY HALL

*The Victorian entry hall served as the transitional space* between the public, outside world and the private, family world. A visitor waited in the hall without intruding on the family.

Nineteenth-century author Edith Wharton characterized the entry hall as "the introduction to the living-rooms of the house…. The hall is a means of access to all the rooms on each floor; on the ground floor it usually leads to the chief living-rooms of the house as well as to the vestibule and street…so that it is the center upon which every part of the house directly or indirectly opens."

The first room a visitor sees, the hall is designed to impress by expressing the homeowner's status and beliefs. Take Margaret Brown's entry hall: Anaglypta, an embossed paper, was a common wall finish, yet the gold paint was a more expensive way to emphasize the room's rich woods. The highly polished woods of the staircase and wainscoting reflected the new technologies of electric lighting and machine carving. The carefully decorated "Turkish corner" showed Margaret's knowledge of current decorating techniques: Orientalism was widely popular, and women's periodicals demonstrated how to create a "cozy corner" such as this one.

The blackamoor statues also reflect the Oriental style. Note that one holds a calling card tray. A telling symbol of the Victorian woman's world, the tray commonly sat on a table or pedestal. The daytime activity of "calling" encompassed both etiquette and class status. A visitor left a small card on the tray during visiting hours. The woman of the house returned the call, or she ignored it—a social snub. A woman could thus communicate with her peers, or even those outside her circle, without actually meeting with them.

A tray for calling cards, a place for your hat and umbrella, a chair to wait in, and the symbolic hearth: These conveyed, in a glance, the homeowner's wealth and worldliness.

*Below:* The entry hall, circa 1910. Margaret updated the room by removing the "Turkish corner" and replacing it with a settee. Ferns and palms decorate the room for a party. Courtesy the Colorado Historical Society.

*Left:* This oil by Charles Alexandre de la Fosse Coëssin depicts a young woman holding a parrot in a boudoir. Margaret may have bought the painting while traveling in France in 1896. A lover of art, she decorated her home with many souvenirs of her travels.

### Restoration Note

When the coved false ceiling was removed, parts of the frieze were found intact. Those portions served as a model for the frieze's restoration.

*four*

# THE PARLOR

*Good breeding consists in concealing how much we think of ourselves
and how little we think of the other person.* ~ Mark Twain

From the entrance hall, visitors were shown into the parlor. The "best" room in the house, the parlor is where the Browns displayed their values and aspirations with physical artifacts. Through such customs as calling, tea drinking, and music, Margaret entertained her most important guests and family members here.

Incorporating a mix of popular eclectic styles—including Colonial Revival floral swags along the frieze and rose-colored wallpaper—Margaret furnished the room with overstuffed Turkish chairs, a polar bear rug, and a piano. Renaissance Revival window treatments showed off the stained glass.

Artifacts like houseplants, a family Bible, and needlework typified the domesticity of the parlor. The traditional parlor suite was symbolic of the family hierarchy: the male in the "head chair," the woman in the next-best chair in the house. A man could sit back and relax as the head of the household, while the woman sat upright, ready to serve her family's needs.

Parlor furnishings also signaled an interest in the cosmopolitan—the notion of being at home in the world. Travel portfolios, pianos or organs, and photographs conveyed knowledge of the world and of high culture. In fact, the parlor expressed two worlds: the domestic, female world and the outside, cosmopolitan world. Parlor furnishings spoke of what it meant to be civilized. Refinement meant progress, and progress meant possessions reflecting accomplishment. The parlor spoke of the family's status but also of the feminine world that women were charged with creating.

The height of the formal parlor was the 1890s, when Margaret bought this home. By 1910, advice manuals promoted a less formal parlor space and talked more of "living rooms." Margaret, however, kept her formalized parlor, perhaps because it displayed her rise to the upper middle class and her own Victorian sensibilities. In her parlor, Margaret showcased her travel souvenirs, art, music, and fashionable tastes—creating a decidedly feminine world within her home.

**Below:** The parlor, circa 1910. Margaret hired a photographer to capture the look of her home as she readied it for a garden party. Courtesy the Colorado Historical Society.

*If the truth be known, all affectations and pretense aside,*

*the dinner, the work over, is the symbol of a people's civilization.*

Robert Laird Collier, *English Home Life*, 1886

### Restoration Note

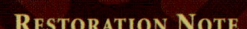

Using surviving plaster medallions beside the fireplace as models, the museum repaired and replaced the broken and damaged plaster dado around the rest of the room.

# THE DINING ROOM

$\mathcal{S}$econd only to the parlor, the dining room was the house's most important entertaining space.

Here, Margaret held luncheons and dinners for family and society. The architectural elements emphasize the room's lofty status. The unusual decorative ceiling—painted to resemble a palm-filled, sunlight-drenched conservatory—serves as a focal point. Bringing the natural world into the home was a popular Victorian practice, as evidenced by the mounted animals and the plant life in other rooms.

Pocket-doors—pairs of doors that slide into the walls—enabled the Browns to close off the room from the others. The doors would remain closed until dinner was served, heightening the excitement of the meal to come. They also served a practical function, maintaining heat during winter months. An outside door leads to the front porch, which served as an extension of the parlor in the summer. Diners could enter through this door once dinner was announced.

The room includes another fine example of Victorian decorating. A tastemaker of the period, Charles Eastlake, dictated the three-part system of wall finishes: a "dado," or the section below the chair rail; a frieze or cornice at the top; and the "field" between. The dado in the Molly Brown House is of combed plaster medallions in metallic paints; the frieze is also plaster. The chair rail is a practical touch: It protects the walls from the furniture.

Other decorative elements, such as the Browns' tapestries on either side of the china cabinet and over the buffet, showcase the room's importance as a space for entertaining. The tapestries' feast and hunt scenes express the themes of bounty and abundance.

*Above:* The dining room, circa 1900. Courtesy the Colorado Historical Society.

MRS. JAMES J. BROWN WAS HOSTESS AT A BEAUTIFULLY ARRANGED LUNCHEON WEDNESDAY. THE INVITATIONS FOLLOWED THE FORM WHICH HAS OF LATE COME INTO VOGUE IN THE EAST AND BY WHICH THE GUESTS ARE BIDDEN TO A LUNCHEON COMPANY, THEREBY INDICATING A SOMEWHAT INFORMAL ENTERTAINMENT. MRS. BROWN'S LUNCHEON, HOWEVER, WAS VERY ELABORATE. THE TABLE DECORATIONS WERE HANDSOME AND WERE IN THE GREEN AND WHITE WHICH IS SO FAVORED AT EASTER TIME. THE TABLE WAS SPREAD WITH THE EXQUISITE LINEN WHICH MRS. BROWN BROUGHT FROM JAPAN. FOR A CENTER PIECE WAS A LARGE WHEEL-BARROW MADE OF EASTER LILIES AND FILLED WITH SIMULATED DUCKS AND CHICKENS. CALLA LILIES WERE SCATTERED ON THE TABLE CLOTH. ~ *1909 NEWS ACCOUNT*

# THE DINING ROOM

*After a good dinner, one can forgive anybody, even one's own relations.* ~ Oscar Wilde

Wealthy Victorians took eating and entertaining to the highest levels. They considered the dining room a venue for educating and civilizing the family. The table was an arena where the family could express its status, and food was a vehicle by which a woman could directly impact her family's health and happiness. The room in which a meal was held, the way the table was set, the people present and where they sat—even the time of day, the day of the week, or the time of the year—all dictated specific (and clearly understood) behaviors.

The late nineteenth century was a time of tremendous social change and immigration, and the upper class felt threatened by the burgeoning lower classes. They developed a code of etiquette that set them apart. In the elaborate table rituals that developed, even the utensils took on specific functions and meanings.

Mass-produced goods enabled middle-class families to acquire a semblance of the prosperity the wealthiest Americans enjoyed. Sears catalogs for 1908 list dinner sets of 56 to 101 pieces, with prices ranging from $3.69 for fifty-six pieces of plain white china to $8.75 for an eighty-piece set of Haviland china.

*Below:* Margaret Brown in the dining room, circa 1900.
This is the only known photo of Margaret inside her home. Courtesy the Colorado Historical Society.

eight

# THE LIBRARY

The Molly Brown House Museum presents the library as it would have appeared in 1910. Before that, the room served as a family parlor. Victorian homes of this size often had more than one parlor—the formal parlor and another for every day.

### RESTORATION NOTE

The bookcases were removed and sold to a family who rented the home while the Browns traveled. The museum got the bookcases back in 1994—76 years after they left the house.

# THE LIBRARY

Margaret turned this room into a library by removing wallpaper on the ceiling and walls and installing bookcases. By placing the family parlor in the rear of the house, she made this space the more public one. And by making this space a library, not only was she following popular decorative styles of the time, but she was making a statement: The dedication of a room to books and reading meant that education and learning took a high priority in her home.

Victorian tastemakers dictated that libraries be furnished to make one "feel at home." Books were the primary decorative elements.

*Above left:* The library, circa 1910; courtesy the Denver Public Library, Western History Collection. *Right:* Margaret's parents strike a somber pose prior to the room's conversion; courtesy the Colorado Historical Society. *Below right:* "Phrenology" held that the skull's contours revealed one's character; heads like this one aided in mapping personality traits. *Facing page:* A souvenir image of the 1893 World's Columbian Exposition in Chicago, where Margaret and J.J. traveled to see the exhibits and "model city" buildings. A circa-1910 postcard depicts a boating and picnicking jaunt.

Books of the period had luxurious bindings that spoke of their value as objects of art and culture. As Edith Wharton put it, "plain shelves filled with good editions in good bindings are more truly decorative than ornate bookcases lined with tawdry books."

Furnished in a simpler style than the parlor or dining room, the library has tobacco-brown walls that showcase the mahogany woodwork, oak bookcases, and the rich, colorful bindings of the books. Travel souvenirs and historical artwork complement the room's intellectual atmosphere.

$\mathscr{S}$ometime between 1904 and 1910, Margaret situated the family parlor at the rear of the home, making it a more private, personal space. The only access was from the library or back porch, not from the formal entrance hall—indicating that visitors were not received in this room.

Traditionally, family parlors were home to everyday activities. School-aged children would do their homework here while the rest of the family might be writing letters, reading, or studying the Bible. Families played parlor games like charades and authors after dinner. In the age before electronic media, Victorians relied on print media and photographs to provide them with views of the world. Stereoscopes—devices that made photographs look three-dimensional—were common in middle-class homes, and stereoscopic "views" could be bought for little money.

We do not know exactly how Margaret furnished this room. But typically, older or cast-off styles of furniture went into the family parlor, as did personal artifacts like family photos, postcards, and travel pamphlets. This was an informal room—you did not have to showcase your best items here. This was a space for relaxing and enjoying solace from the pressures of the outside world.

## THE SUNROOM

𝒱𝒾𝒸𝓉𝑜𝓇𝒾𝒶𝓃 𝒽𝑜𝓂𝑒𝑜𝓌𝓃𝑒𝓇𝓈 𝒹𝑒𝓈𝒾𝑔𝓃𝑒𝒹 the first floor of their home to impress visitors, and to give those guests a taste of the owner's public persona. But the second floor was usually the homeowner's private space.

The second floor of Margaret Brown's home included a sitting room on the west side. The room's proximity to both the entrance hall and the bedrooms made it a transitional space between the public and private worlds. You could watch or listen to the activities below, without actually being present. Close friends could visit without the formality of the front parlor and, at the same time, without intruding into the more private spaces of the bedrooms.

Margaret's sunroom had an added appeal: It looked out onto a spectacular view of the Rocky Mountains (a view lost with the construction of the high-rise across the street). A door leads to a balcony, where hired musicians entertained Margaret's guests.

## THE BEDROOMS

*Lang designed the Molly Brown House* with all five sleeping chambers on the second floor. Bedrooms were multipurpose rooms—home not just to sleeping but to dressing, letter writing, reading, and sewing as well.

*…She entered her bedroom, with its softly-shaded lights, her lace dressing-gown lying across the silken bedspread, her little embroidered slippers before the fire, a vase of carnations filling the air with perfume, and the last novels and magazines lying uncut on a table beside the reading-lamp…*
~ Edith Wharton, *The House of Mirth*

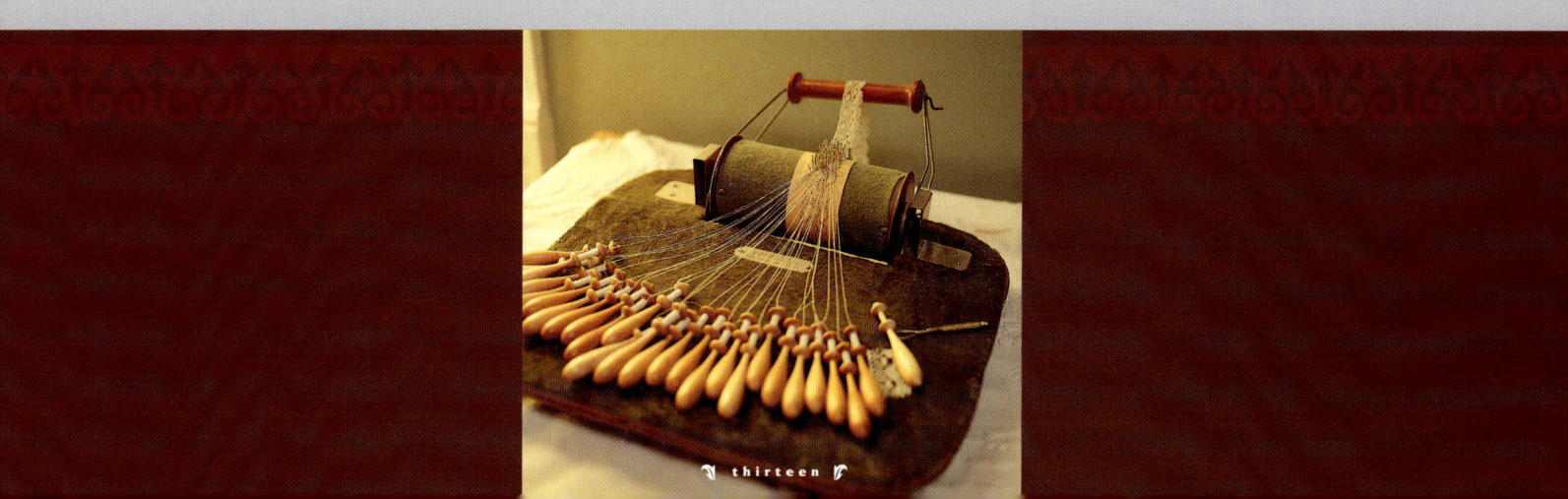

thirteen

*Top:* The large south bedroom features a bay window and intricate decorative ceiling. The day bed was for afternoon naps.

*Center:* The guest bedroom, or possibly daughter Helen's room.

*Bottom:* The largest bedroom was also the only one with a fireplace.

## THE BEDROOMS

𝓑𝓮𝓭𝓻𝓸𝓸𝓶𝓼 𝓪𝓵𝓵𝓸𝔀𝓮𝓭 𝓽𝓱𝓮𝓲𝓻 𝓸𝓬𝓬𝓾𝓹𝓪𝓷𝓽𝓼 𝓽𝓸 𝓮𝔁𝓹𝓻𝓮𝓼𝓼 𝓽𝓱𝓮𝓲𝓻 𝓸𝔀𝓷 𝓼𝓽𝔂𝓵𝓮𝓼. In 1907 *Harper's Bazaar* wrote: "Every opportunity should be given for the development of individuality in a room which is preeminently the corner of the home which is truly home to the occupant, where the taste of no one, either guest or relative, need be considered."

Like the family parlor, bedrooms inspired a mix of new and old styles of furniture, carpeting, and window treatments. Bedroom suites generally included a wood or iron frame bed, dressing table, and washstand. (Reformers touted the iron frame for its ease of cleaning.) In upper-class households, men and women slept apart.

Dressing took on added meaning to the Victorian woman, who changed her clothing based on the activity. Calling, receiving visitors, shopping, bicycling, attending formal balls—all required a specific form of dress. Hygiene was of the utmost importance. Research into the causes of disease produced a number of manuals on cleanliness, and dresser sets came with items for hair, nail, complexion, and tooth care.

Although her house had a telephone, Margaret still relied on notes for invitations, correspondence to friends and business associates, thank-you notes for luncheons, and menus for the cook. These she wrote at her desk, in the privacy of her bedroom.

### Restoration Note

Evidence showed that Margaret decorated three of the bedrooms in a mix of revival styles with silk brocade walls, cornices over the windows, and painted trim and woodwork. Lighter colors like pale blue and light green gave the rooms an airy, open feel.

# THE STUDY

Once another bedroom, the study functioned as a private "den" or smoking room. Spaces like this were quite popular with Victorian men. Outfitted with a roll-top desk and chaise lounge, the room offered a retreat from the downstairs formality and a haven for conducting personal correspondence.

## Restoration Note

A sample of the room's wallpaper—which is of Oriental or Turkish influence—turned up during the restoration. It was big enough that the pattern could be reproduced

ℬy the turn of the century, most middle-class homes boasted an indoor bathroom. Fixtures consisted of a porcelain tub, a sink, and a toilet with water closet. No more chamber pots or trips to the outhouse!

Bathrooms were meant to be practical and sanitary, thus not heavily decorated. Most appropriate were glazed ceramic tiles in white, buff, or gray. Ads stressed the sanitary element, touting that "you cannot have too many safeguards for the health of your family and self.…Better by far [to] pay for good plumbing than for doctor bills resulting from defective sanitary equipment."

Having the most up-to-date bathroom symbolized that you owned a modern home—an emblem of prosperity.

## THE KITCHEN

The Molly Brown House kitchen is laid out in functional fashion: stove, sink, and cookware all easily accessible. Advice manuals of Margaret's day advocated the necessity of a clean kitchen in efficiently performing the household chores.

Cupboards for storing china, linens, and silverware line one wall. The cook's pantry holds seasonal goods, canned items, and cookware. The butler's pantry served as the buffer zone between dining room and kitchen. Guests could see into it when the door was open, so it had to be kept clean and well decorated.

The Victorian kitchen was the domain of the domestic help. Margaret employed two Irish maids and a housekeeper who doubled as cook. The female employees lived on the third floor, while the male butler or carriage boy lived above the carriage house. The baker, butcher, grocer, iceman, and coal company all delivered.

Margaret's kitchen boasts the technological advances of her day. The city water system brought water to the home. Electric light allowed the cook to work with more than just natural daylight. Electricity also ran the annunciator, or call box, that rang from throughout the house. The coal stove, hot and cold running water, and an icebox enabled the cook to create all the family's and servants' meals in one place.

Gadgets like grinders, apple peelers, cherry pitters, and biscuit makers epitomize the domestic revolution of the late nineteenth century. The electric washing machine, electric iron, and eventually the vacuum arrived in the early 1900s. Thus, technology didn't just dictate the functions of rooms; it helped make the house run.

*Above left:* The annunciator, or servant's call box. Arrows directed the maids to the caller's location.

### Restoration Note

Museum staff relocated the servants' staircase when wall demolition revealed the top riser. A month later, they discovered Lang's original drawings—showing that the cabinetry and stairwell were now situated exactly as Lang had specified.

*…As she remarked to her pupil, a good cook was the best introduction to society.*
~ Edith Wharton, *The House of Mirth*

# THE CARRIAGE HOUSE

*The carriage house sheltered the transportation devices* that were so vital to the Browns' modern lifestyle. Like most upper- and middle-class Denverites, the Browns rode a carriage and, later, Margaret owned a Fritchle 100-mile electric car. Fritchle built his electric cars just a few blocks from here at what later became known as "Mammoth Gardens"—a building at Colfax Avenue and Clarkson Street that still stands today.

On the second floor of the carriage house slept the groomsman, who maintained the horses and carriage. The Browns stored hay in the loft, and their horses lived here as well.

Designed by William Lang to complement the home, the carriage house went up a few years later. It maintains the same eclectic lines as the home, not to mention the play of rhyolite and sandstone that give the main house so much of its character.

In rehabilitating the carriage house for use as the Museum Store and offices, the museum has kept all changes sensitive to the structure's original architecture. The sliding stable doors, horse compartment windows, hay door, and open beams all remain in view while at the same time allowing for the carriage house's newfound uses.

The Molly Brown House and its carriage house form a truly elegant pair. In their mix of locally quarried stonework and their lively Queen Anne and rough-hewn Richardsonian Romanesque designs, they stylishly reflect the prosperity of a booming Victorian-era Denver.

*Facing page:* Larry and Helen Brown enjoy a pony cart ride, circa 1895; courtesy the Colorado Historical Society. Margaret Brown on horseback around the same time; courtesy the Colorado Historical Society.
*Below:* Photo courtesy Roger Whitacre.

*The Molly Brown House Museum* Using educational programs and artifact acquisition, the Molly Brown House Museum interprets Margaret "Molly" Tobin Brown's life—primarily between the years 1894 and 1912—in order that a broad public and future generations may understand and appreciate the social, economic, and political aspects of Victorian life in Denver. The museum, a property of Historic Denver, Inc., preserves Margaret Brown's historically significant home through continued restoration.

The museum researches, collects information, and acts as a resource for any and all interested in Margaret Brown's entire life (1867–1932). The museum also researches, collects information, and acts as a resource for those interested in the *Titanic* disaster, specifically Margaret Brown's role in that tragedy.

*Historic Denver, Inc.* Historic Denver, Inc., is a citizen organization whose purpose is to preserve Denver's significant historic fabric, its distinctive architecture, and its cultural landscapes, which are the tangible reflections of our heritage and the foundation of our quality of life.

Our responsibility as a nonprofit corporation is to be a catalyst for and advocate of ideas, programs, actions, and plans which enable our community to respect and carry forward the preservation of this heritage.

Elizabeth Owen Walker, a Denver native, is the curator of the Molly Brown House Museum. She holds an M.A. in American studies from George Washington University.

Jeff Padrick owns and operates KLUG Studio in Denver. He specializes in commercial, architectural, and still-life photography.

Violet Carlon, a Denver-based graphic designer, specializes in corporate and museum communications.